Matter
Solids, Liquids, and Gases

Mir Tamim Ansary

RIGBY
INTERACTIVE
LIBRARY

This edition © 1997 Rigby Education
Published by Rigby Interactive Library,
an imprint of Rigby Education,
division of Reed Elsevier, Inc.
500 Coventry Lane
Crystal Lake, IL 60014

Printed in the United States

00 99 98 97 96
10 9 8 7 6 5 4 3 2 1

Library of Congress Cataloging-in-Publication Data
Ansary, Mir Tamim.
    Matter: solids, liquids, and gases / Mir Tamim Ansary.
      p. cm. — (Science all around me)
    Includes bibliographical references and index.
    Summary: Explains the basic properties of matter through looking at everyday experiences and direct observation.
    ISBN 1-57572-110-4
    1. Matter—Juvenile literature. [1. Matter—Properties.]
I. Title. II. Series.
QC173.36.A57 1996
530.4—dc20
                                         96-22988
                                           CIP
                                         AC

Cover designed by Lisa Buckley
Interior designed by Jean Wheeler
Commissioned photography by Sharon Hoogstraten and Zul Mukhida
Consultant: Hazel Grice

Acknowledgments
The publisher would like to thank the following for permission to reproduce photographs. Eye Ubiquitous, p. 5; Tony Stone Images (Paul Chesley), p.20.
Every effort has been made to contact copyright holders of any material reproduced in this book.
Any omissions will be rectified in subsequent printings if notice is given to the publisher.

> **Note to the Reader**
>
> Some words in this book are printed in **bold type**. This indicates that the word is listed in the glossary on page 24. This glossary gives a brief explanation of words that may be new to you and tells you the page on which each word first appears.

Visit Rigby's Education Station® on the World Wide Web at http://www.rigby.com

MAR - - 1998

Contents

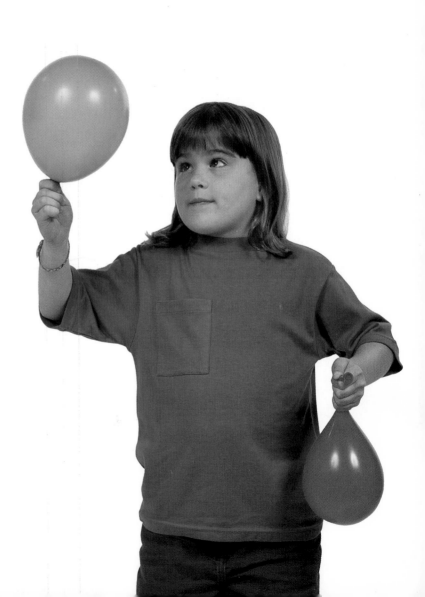

What Is Matter?

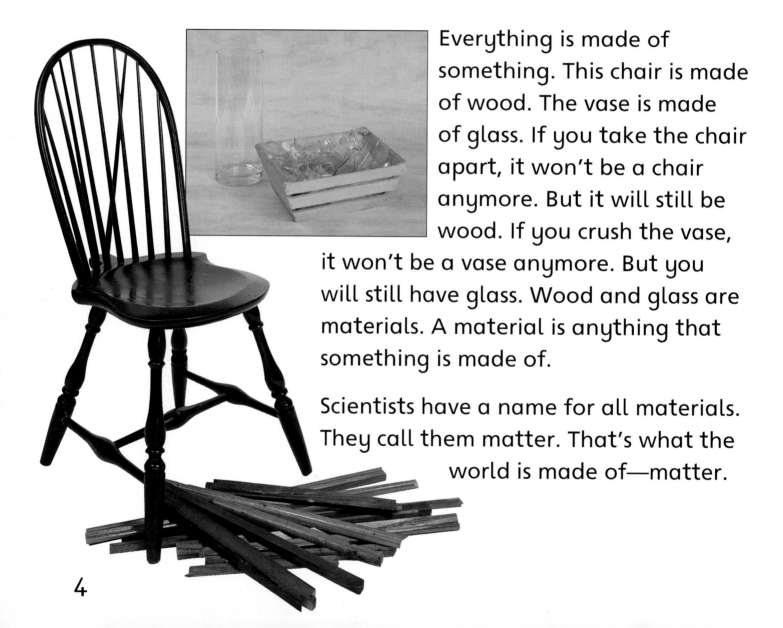

Everything is made of something. This chair is made of wood. The vase is made of glass. If you take the chair apart, it won't be a chair anymore. But it will still be wood. If you crush the vase, it won't be a vase anymore. But you will still have glass. Wood and glass are materials. A material is anything that something is made of.

Scientists have a name for all materials. They call them matter. That's what the world is made of—matter.

See for Yourself . . .

Different things can be made of the same material.

- Find three things made of wood in your house.

- Find three things made of glass in your house.

- Find one other material in your house or yard.

How are the things made of the same material different from each other?

5

Size

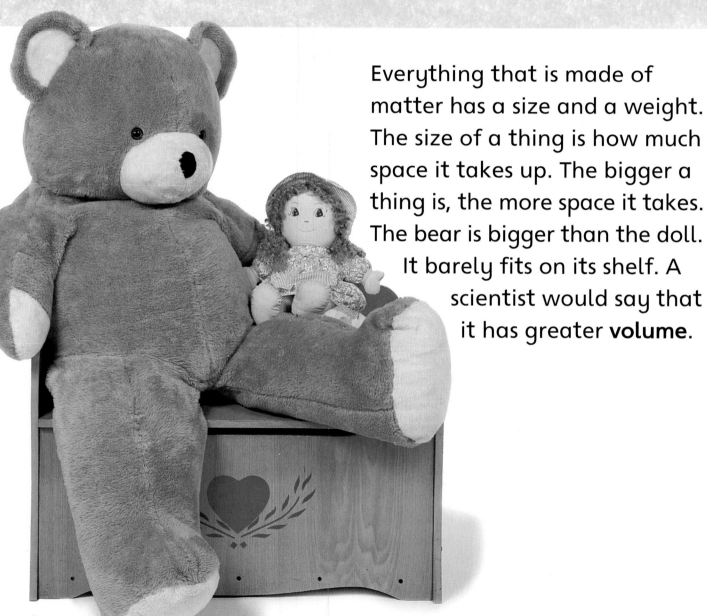

Everything that is made of matter has a size and a weight. The size of a thing is how much space it takes up. The bigger a thing is, the more space it takes. The bear is bigger than the doll. It barely fits on its shelf. A scientist would say that it has greater **volume**.

See for Yourself . . .

Size is something you can see. Take a look around your house.

- Find something smaller than your head.

- Find something bigger than your head.

- Find something about the same size as your head.

Weight

Every material has weight. The marble weighs just a tiny bit. The bowling ball weighs a lot more. It is heavier than the marble.

You cannot see weight. But you can feel it when you try to lift something.

See for Yourself . . .

- Find something around your house you can lift easily. Then find something that feels heavier. Weigh both objects. Does the second object weigh more than the first?

- Close your eyes. Have a friend set several objects in front of you. With your eyes still closed, lift an object. Try to pick the lightest and the heaviest objects.

- Now try it with your eyes open. Lift an object. Then try to pick another object with the same weight. Weigh both objects. Were you correct? Do they weigh about the same?

Density

Size and weight do not go together. Something small can be heavy. Something big can be light. Two things that are the same size can have different weights.

The brick and the bread are about the same size. But the brick is much heavier. It is made of denser material. Density is the amount of matter packed into a space. Think of a crowded elevator. As more people push inside, the crowd gets denser.

See for Yourself . . .

- Get two plastic cups.

- Fill one with liquid cream and one with whipped cream.

- Weigh both cups.

The cup with cream is heavier. That's because cream is denser than whipped cream.

Three States of Matter

Water is soft and wet. But if you boil it, water turns into **steam**. Then it seems to disappear. If you freeze water, it turns into ice.

Steam, water, and ice are made of the same material. When water freezes, it becomes a *solid*. When it melts, it becomes a *liquid*. When it boils away, it becomes a *gas*.

Solid, liquid, and gas are the three **states** of matter. All materials can be sorted into these groups. All three are present in Tony's glass of root beer. The glass is solid. The root beer is liquid. The bubbles in the root beer are filled with gas.

See for Yourself . . .

Solids, liquids, and gases—they are all around us.

- Touch three different solids.

- Name two liquids that you can drink.

- Squeeze a ball. There is a gas inside it.

Solids

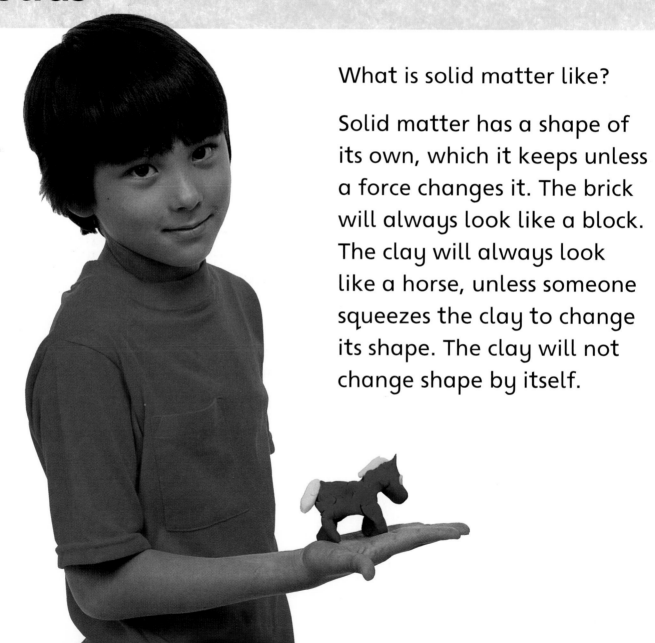

What is solid matter like?

Solid matter has a shape of its own, which it keeps unless a force changes it. The brick will always look like a block. The clay will always look like a horse, unless someone squeezes the clay to change its shape. The clay will not change shape by itself.

14

See for Yourself . . .

Solids do not flow together when they touch.

- Put some of your socks in a drawer.

- Put someone else's socks in the drawer.

Mix them up. Look closely. Can you still see your socks? Can you still find the other person's socks? The two sets of socks stayed apart because they are solids.

Liquids

Liquids can be poured. Thin liquids, like this juice, pour quickly. Thick liquids, like honey, pour slowly.

When does one plus one equal one? When you add drops of water. Liquids of the same kind flow together when they touch. Oil mixes with oil. Water mixes with water. Add some food coloring to water and watch them mix.

See for Yourself . . .

A liquid does not have a shape of its own. It takes the shape of its container.

- Pour water into a glass. What shape does the water have?

- Pour the same water into a bottle. What shape does it have now?

As the water moves to a new container, only its shape changes. The volume stays the same. Each container in the picture is holding one cup of water.

Gases

Hardly any space on earth is really empty. Spaces that look empty are usually filled with air. Air is a mixture of gases.

A gas has no definite shape or volume. It spreads out to fill the space it's in. If the space becomes smaller, the gas packs together and becomes denser.

You can blow air into a beach ball. No matter how hard you squeeze, you can't squash the ball.

See for Yourself . . .

Gases are lighter than liquids or solids.

- Fill a balloon with water.

- Fill another balloon with your breath.

- Hold a balloon in each hand. Which is heavier?

Usually you can't see gases. But you can feel them. Blow on your hand. What you feel is air pressing against your skin.

Heating Matter: From Solid to Gas

Many solids can change form. It doesn't take magic. All it takes is **heat**. Heat turns many solids into liquids. It turns most liquids into gases.

Enough heat can even melt a rock. There is hot liquid rock inside the earth. You can see some of that liquid pouring out of this volcano. As it cools, the liquid will turn into solid rock.

See for Yourself . . .

You can turn a solid into a liquid.

- Put some solid ice cubes in a bowl.

- Set the bowl in a warm place.

- Watch the solid ice slowly turn into liquid water.

In a few days, the water will turn into a gas called water **vapor**. You can't see this gas. It mixes into the air.

Cooling: From Gas to Solid

Matter changes when you cool it, too. When water vapor in the air touches a cold **surface**, such as the cold ground, it cools and turns from a gas into a liquid. This creates dew, the glistening beads of water you find on the grass on some mornings.

See for Yourself . . .

- Put ice water in a glass and dry off the outside of the glass.

- Set the glass in a warm place.

- Wait 20 minutes.

- Now feel the glass. Where did the water come from?

Water vapor in the air touched the cold glass. As the vapor cooled, it turned from a gas into liquid—the water on the outside of the glass. If the water were to freeze, it would turn into ice, a solid.

Glossary

heat High temperature 20

states Conditions or forms of something 12

steam Water in the form of hot gas 12

vapor Tiny bits of something in the air 21

volume Amount of space something takes up 6

Index

Further Readings

Evans, David and Claudette Williams. *Building Things.* Dorling Kindersley: New York, 1992.

Evans, David and Claudette Williams. *Make It Change.* Dorling Kindersley: New York, 1993.

Peacock, Graham. *Materials.* Thomson Learning, 1995.